ROYAL
OBSERVATORY
GREENWICH

T0136546

Black Holes

Dr Ed Bloomer

Royal Observatory Greenwich
Illuminates

First published in 2021 by Royal Museums Greenwich,
Park Row, Greenwich, London, SE10 9NF

ISBN: 978-1-906367-85-5

At the heart of the UNESCO World Heritage Site of
Maritime Greenwich are the four world-class attractions
of Royal Museums Greenwich – the National Maritime
Museum, the Royal Observatory, the Queen's House and
Cutty Sark.

rmg.co.uk

A CIP catalogue record for this book is available from
the British Library.

Typesetting by ePub KNOWHOW
Cover design by Ocky Murray
Diagrams by Dave Saunders
Printed and bound in the UK by CPI Group (UK) Ltd,
Croydon, CR0 4YY

MIX
Paper from
responsible sources
FSC® C020471

About the Author

Dr Ed Bloomer is an astronomer and science communicator. Prior to joining Royal Observatory Greenwich, Ed completed his PhD at the Institute for Gravitational Research at the University of Glasgow, where he primarily tried to work out what happened when small black holes spiralled into bigger ones ...

Entrance to the Royal Observatory, Greenwich, about 1860.

About Royal Observatory Greenwich

The historic Royal Observatory has stood atop Greenwich Hill since 1675 and documents over 800 years of astronomical observation and timekeeping. It is truly the home of space and time, with the world-famous Greenwich Meridian Line, awe-inspiring astronomy and the Peter Harrison Planetarium. The Royal Observatory is the perfect place to explore the Universe with the help of our very own team of astronomers. Find out more about the site, book a planetarium show, or join one of our workshops or courses online at rmg.co.uk.

Contents

A Brief Introduction

After almost any public astronomy lecture someone will ask about black holes, regardless of whether the lecture mentioned them or not. In terms of public engagement with scientific concepts, this is great news. They're a fascinating fusion of the simple and the complex, a demonstration of the physics we understand and a test-bed for more that we do not. I love talking about them! Nevertheless, people often have all sorts of strange ideas about what they are and how they work. Lots of misinformation is flying around, helped along by sensationalist articles and no small amount of misleading science fiction.

One way to deal with this would be to sign up everyone for some astrophysics courses and reach the subject of interest through a firm foundation of mechanics, mathematics and a few other disciplines. But frankly we'd lose most people along the way and I think we make more progress discussing concepts rather than proposing a mathematical treatment of 'problems'. What follows then is my guide to what black holes are and how they relate to some other areas of science, a little bit about how we came to understand some of those issues and a few consequences of our explanations. Although I can't promise it will be absolutely comprehensive, I think we'll be able to cover these ideas pretty well without the need for a full lecture course.[1]

---

[1] Since this isn't an astrophysics course, there are lots of things we might have to skip past or only mention in passing. Every now and then, I'll chime in with extra little pieces if I think they might be useful, weird or just fun! They're quite frequent because I get distracted easily ...

In fact, I'm pretty sure we won't even need a calculator. Probably.

Nevertheless, some sort of starting definition would be useful before we start our conceptual deep diving. What are black holes anyway?

Well, simply put, black holes are objects that generate such extreme gravitational pull that not even light can escape them. They are astronomical objects, often the result of the death of massive stars, and can in theory devour anything that gets close enough. They only have a few physical properties, so in some sense are easy for astronomers to describe mathematically, but the result of their interactions with the rest of the Universe are extremely complicated – to the point that we do not fully understand them. Further, the mechanics of black holes and their interactions with the rest of the Universe are … well, they're strange. They don't always make sense the first time you read about them.

It gets worse before it gets better. If you were to repeat any part of the previous paragraph to a group of astronomers, you would be met with a chorus of 'well, *sort of* …', and 'hmm, I suppose it depends what you mean by …' and 'perhaps from a certain point of view …'. All of which, I understand, is very unsatisfactory and even intimidating to a new reader. Why can't we just explain everything in a concise, straightforward manner?

I hope you'll trust me. Delving into the strangeness is going to be fun and I think you'll come to appreciate the responses from our imaginary group of astronomers. Fair warning: you may end up rolling your eyes at a lot more science-fiction depictions of black holes.

If this seems daunting, don't worry. We're going to set things up with straightforward ideas and build on them. At first, some of these might seem rather tangential, as if we're avoiding the main issue, but once we establish some basic principles we'll see how twisted they become once we consider the extreme conditions represented by black holes. The strangeness will become the logical conclusions of those principles.

Ahem.

Obviously, the best way to start is to jump straight into explaining the underlying framework of the entire Universe …

Setting the Scene

Newtonian physics

Unfortunately, the history of science is not a linear progression from ignorance to enlightenment. Some developments force us to reassess the foundations on which they were built and find them completely inadequate. Other ideas are very accurate in everyday situations but won't hold true in the extreme circumstances involving black holes, so we're going to have to try and set the scene a little.

Most of us, consciously or not, go about our lives as if we live in a Universe governed by the kind of natural laws that

Isaac Newton (1643–1727) studied in the 17th and 18th centuries. A thousand pages and more could be written about science during Newton's lifetime and the enormous contributions he made that have ensured his relevance so long after his death. However, we're going to have to narrow it down, and gravity is probably what springs to mind at first. It's certainly important! I'll skip right past whether or not Newton was inspired by a falling apple (let alone whether it fell on his head). A little reminder, though, that Newton's theory of gravitation refers to his framework for explaining the observed effects of gravity.[2] In brief, **matter** generates an attractive gravitational field, and **masses**

---

[2] Theories aren't just an idea that someone has had. The theory of evolution is not the notion that something called evolution exists, but the framework that formalises the observations we've made about evolutionary changes and makes predictions about changes we should observe in other circumstances.

in this field will have a resulting **force** applied to them, which causes them to accelerate. If you double the force, you double the acceleration (or double the force and ensure the same acceleration for twice the mass).

As it turns out, Newton's theories of gravitation and mechanics aren't quite ... well, there. They don't exactly work. A Newtonian framework is a useful approximation for things in lots of circumstances (sometimes very useful), but it starts to fall apart when you consider particularly massive objects or very high velocities. Black holes involve both situations, so Newton alone won't suffice.

Let's not move on too quickly though. This Newtonian universe is a useful way to start considering some things, even if it isn't strictly 'true' according to more recent advances.

Newton was a proponent of the **corpuscular theory of light,** which considered light to be made of little particles, or

corpuscles. These were not the **photons** that we think of today, but his theory was a useful step along the road to our current understanding and, if we keep it in mind, corpuscles can get us closer to the 'blackness' of black holes.

Imagine holding something in your hand and accidentally dropping it. As soon as you let it go, gravity pulled it towards the ground. A simple thing really, at least at first glance. However, there's actually quite a lot going on.

In your hand, that object had **potential energy**, a measure of its energy in relation to other objects.[3] At the start though, it had no **kinetic energy**, a measure dependent on its mass and **velocity**, because you were holding it stationary. When released, the force of gravity caused the object to accelerate, and the potential

---

[3] The Earth, most importantly, assuming you dropped your object on this planet.

9

energy started to be converted into kinetic energy (see Figure 1). Basically, the object started higher up and static and ended up dropping and hitting the floor at speed.

There are lots of complicating factors, but what's important is the change in energy. Drop your object from higher up and you'll convert more potential energy

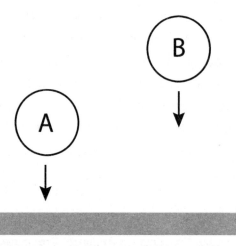

Figure 1. Dropped from rest, both A and B will accelerate under the force of gravity and fall to the floor. B, starting at a higher point, has greater potential energy, which is converted into greater kinetic energy. B will hit the floor at a higher velocity than A.

into kinetic energy, and your object will hit the floor at a higher velocity.

Interesting thing number one: the gravitational force pulling the object downwards depends on its mass, the mass of the attracting body (the Earth) and the distance between them (mathematically, the square of this distance, but let's not worry about that). So, a more massive object – not bigger, I really mean something with more mass – 'feels' more gravitational attraction and the net result is the same acceleration.[4] That's why it shouldn't matter whether you accidentally dropped a cup, a plate, a bowling ball or a feather.

---

[4] When we talk about massive objects, we mean things that have mass, when colloquially we might mean 'big'. An elephant is massive because it has mass, but so is a mouse. Admittedly, the elephant is more massive, and it also happens to be a lot bigger. But a black hole with the same mass as the entire Earth would only be the size of a large marble (as we'll see later).

Galileo Galilei (1564–1642) demonstrated this by dropping objects of different weights from the Leaning Tower of Pisa, showing that they fell at the same rate, rather than the heavier ones falling faster (see Figure 2).[5] In 1971, David Scott dropped a hammer and a feather on the Moon during the Apollo 15 mission. True to expectation, they hit the Moon's surface at the same time, demonstrating that this is indeed what we hope and expect to be a universal law of attraction. It clearly isn't even the Earth that is special – you can do the experiment in a vacuum chamber on Earth and the results will be the same. Nor is the Moon special, of course. It is the mass of the attracting body that's important.

---

[5] Actually, he probably didn't. People just said he did and it entered into popular culture. But others have done all sorts of similar things, including rolling things down inclined planes to demonstrate the principle.

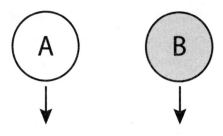

Figure 2. Here, A and B are vastly different masses, experiencing different gravitational attractions. Gravitational acceleration is, however, independent of the masses of A or B. Dropped from rest at the same moment, A and B will hit the floor at the same time.

Interesting thing number two: in our case, 'downwards' actually just means towards the centre of mass of the attracting body. For a sphere like the Earth,[6] the centre of mass is right at the centre of the object. Stand at the North Pole and get a friend to stand at the South Pole and you can imagine that you're standing upside

---

[6] Ok, it's not a perfect sphere, but it's close!

13

down with respect to one another, but really you're both orientated so that your 'down' is towards the centre of the Earth.[7]

Let's look at it the other way. If we gave an object some kinetic energy and aimed it 'up' and away from the centre of the Earth, it would climb out of that gravitational well as best it could, trading kinetic energy for potential energy until the exchange was complete. For a moment in time and at a particular height, it would be stationary, then plummet towards the ground again. This could describe us throwing a bag of sugar into the air. Instead, just think of the moment the bag of sugar leaves your hand. It has a velocity and we're going to see how high that kinetic energy gets it.

_____

[7] 'Up' and 'down' don't hold much currency in many fields of science and particularly in astronomy. 'Towards' and 'away from' can be more useful. Little comfort if you fall off a ladder, but perhaps something to distract you before you hit the ground.

Fireworks and space rockets operate under a different mechanism to thrown objects. We're continuously burning fuel, converting chemical energy into kinetic energy, at least until we run out. When a rocket uses up its fuel and jettisons the fuel tanks, it is in the process of moving and has considerable kinetic energy. But with no more fuel to burn, and caught in the Earth's gravitational field, it will also continue for a while before halting, then plummeting towards the Earth once again.

Now, that particular height the object is able to reach in the air depends on the kinetic energy and the mass of the attracting body. Even ignoring things like the lack of wind resistance, you should be able to throw a bag of sugar higher on the Moon than on Earth, because **gravitational acceleration** on the surface of the Moon (the measure of the attractive force the sugar will 'feel') is only about one-sixth of that felt on the surface of the Earth.

This phenomenon is not limited to the interactions of the Earth and bags of sugar, of course, but is a universal physical law governing the behaviour of massive objects. In our example, there is a mutual attraction between the masses, which means the Earth is being pulled towards the bag of sugar too. It's just that the Earth weighs almost 6,000 yottagrams. Or 6 quadrillion kilograms, if you prefer. In practice, this means there is negligible effect on the Earth in our experiment. But it is genuinely an important point. Mutual attraction is just that. Everything 'notices' everything else. If we start to consider objects of more equal mass, we'll have to remember that.

So, let's pretend we're in an empty universe except for two objects called A and B. And let's say B is much more massive than A. We're going to launch A away from B by giving it an instantaneous boost of kinetic energy, setting its velocity directly away from the centre of mass B.

16

But we don't want to send it 6 feet away, or 100 kilometres away. Let's get it out to infinite distance! Mathematically nice, and practically unrealistic, we'll set it so the conversion from kinetic to potential energy means that A achieves zero speed, infinitely far away from B.

If this seems confusing, let me restate the same thing slightly differently. Imagine you're standing on B, looking at A. One moment, A was resting on the surface of B, completely still. The next, A was imbued with kinetic energy, causing it to fly 'straight up' (that is, away from B's centre of mass). The gravitational attraction of B causes A to bleed kinetic energy, slowing down as the gravitational potential energy builds up. Remember, the gravitational attraction of B lessens the further from A it gets, so the exchange rate of energy changes. In our scenario we've carefully given A just enough energy that it actually

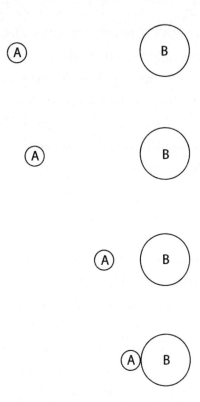

Figure 3. A and B are mutually attracted from rest, but B is so much more massive we're assuming its motion is negligible. A is accelerated towards B and crashes onto its surface. This diagram can be reversed; instantaneously impart the crashing velocity to A (but point it in the opposite direction) and it will move away from B, decelerating until it reaches zero velocity at the position shown at the top. But what happens when the velocity needed to leave exceeds the speed of light?

takes infinite distance for the velocity to reach precisely zero (see Figure 3).

It can be confusing when dealing with abstracts, but I think it emphasises an important point. We're interested in the mass of A and B and I've only mentioned them in comparison to one another. Nothing else is important here: not their colour, size or any other property. You can imagine two very different bags of sugar or launching a spherical mouse from the back of a spherical elephant (we could do it without spheres, but geometrically they're the easiest to deal with). Failing that, imagine Newton avenging himself on an apple by chucking it infinitely far away. Don't worry, we're getting to black holes!

What's important is the launch velocity that allows us to achieve this trade-off. Remember, we're not talking about a rocket launch, burning up fuel to power its way away from B. In our hypothetical scenario we're talking about what we call the **escape velocity**. It is different for objects with different masses – on the Earth it is a little over 11 kilometres per second, but on the Moon it is less than 2.5 kilometres per second.

Basically, under ideal circumstances (again, no air resistance, no being fired off at an angle, no worrying about the gravitational attraction of anything else in the Universe, etc.), if you achieve escape velocity, gravity is not going to keep you on the ground. No need to fly or use up fuel to keep going, you'll be up and away.

The other way to consider the issue is to ask what would happen if you were to drop A onto B from an infinite distance starting with zero velocity. Just the reverse: A will crash into B at this escape velocity. Fair enough, it's travelling in the other direction, but that's just because we've set it up that way.

The escape velocity depends on the mass of attracting bodies, but let's make A a corpuscle of light. It travels at the speed of light and we're imagining that it has a tiny little bit of mass (almost nothing, I suppose, but not actually nothing) and is affected by gravity.

Last step: imagine that B is so massive that the escape velocity is not around 11 kilometres a second like on Earth, or even 100 or 1,000. Imagine that B is so massive that the escape velocity is greater than the speed of light. A has to travel faster than the speed of light to escape. This is impossible! We've hit the speed limit of the Universe.[8] Not even light itself can escape B. It does not radiate anything, it does not reflect light. It is an object of utter darkness. We've imagined a black hole.

---

[8] I realise that putting a speed limit on light itself might seem like cheating, but a proper explanation would be too substantial to include and would form the core of a lengthy discussion of the General Theory of Relativity. You'll have to accept this one, but I'll try to avoid anything relying on 'that's just how the Universe works' from here on.

We're not the first to do so, though our preferred term was not the first one used. In 1784, John Michell (1724–93) proposed 'dark stars' from which corpuscles of light could not escape due to their enormous gravity.[9] Such objects would not be identifiable by sight since they would not be luminous, though we might 'see' them by their gravitational influence on luminous objects nearby. The French polymath Pierre-Simon Laplace (1749–1827) had similar thoughts.

Unfortunately, these ideas didn't gain much traction and not just because the corpuscular theory of light doesn't hold true. Rather, developing this idea of a 'black hole' from a Newtonian framework was so ahead of its time it basically got ignored.

---

[9] Michell was a somewhat obscure polymath who nevertheless made contributions in many fields including physics, and to whom appropriate recognition is coming, slowly!

Einstein's spacetime

Enter Albert Einstein! I'm sure you've heard of him. In the early 20th century, he devised a different framework to explain how the Universe works. The Special Theory of Relativity came first in 1905 and then, a decade later, he produced the General Theory of Relativity, which seemed to encompass … well, just about everything. Farther-reaching than Newton's framework, it seemed like Einstein's ideas could deal with any situation we might imagine. Now, it would be possible to write a very long book or spend your entire career explaining the intricacies of relativity, but for our purposes there are a few things we should pay particular attention to.

The first is that we live in a **spacetime**. Think about how objects can move in three-dimensional space, then add travelling through time. Now you have a four-dimensional spacetime. Travelling through time is like travelling in any direction in three dimensions.[10] Crucially, objects can move through time at different rates relative to an observer. Think of the Universe less as some clockwork construction with a steady rhythm governing everything's movements, and more as flexible, warpable rubber sheets. Or perhaps something more fluid, like an ocean with different densities and flows of liquids.[11]

Keep in mind that an observer is not necessarily a person actively looking. Our observer might be a clock keeping count, for instance. As we'll see later, we want to be free to shift our point of view without worrying how to do so in practice (like when you fall into a black hole).

---

[10] Formally, it is common to think of some combination of orthogonal (that is to say, perpendicular to each other) X, Y and Z vectors. In more simplistic terms, we're talking about any combination you can think of using 'forwards/backwards', 'left/right' and 'up/down'.

[11] If you can think of better metaphors for the structure of the Universe, let someone know!

The second thing to keep in mind is that light is special. Not only is the speed of light the universal speed limit, but in a relativistic sense a photon won't experience any time passing. It isn't that we were worried about photons getting bored or old as they travelled but it does mean we can use them as an interesting tool. With a constant speed of light, we can work out the **proper distance** between events: a distance that includes separation in time as well as space.

When are you going to talk about black holes, though?!

I haven't even got into differential geometry and metric tensor mathematics yet. Oh, all right then.

It does sometimes help to think of spacetime as being a bit like a rubber sheet, which can be deformed by placing massive objects on it. Spacetime has **curvature** and things will move through spacetime differently depending on how curved spacetime is (see Figure 4). Remember, space and time are linked here, so we have a framework within which the velocity of an object affects its passage through time, and objects themselves cause changes to spacetime curvature. An extremely quotable summary comes from the American theoretical physicist John Archibald Wheeler: 'Space tells matter how to move, matter tells space how to curve.'

Even light itself pays attention. We've moved onto the dual particle/wave nature of photons of light (and therein lies its own story). Photons, though, are also affected by the curvature of spacetime and they follow that curvature. But because we have this constant speed of light, photons can actually be used to map out the spacetime they're in. This is extremely useful. If absolutely everything got distorted, how would you know what the distortions were?

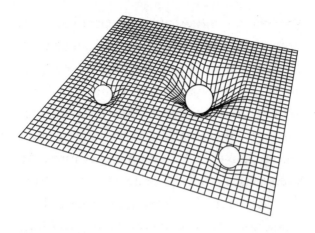

Figure 4. Visualising spacetime curvature is difficult. The classic example is to consider spacetime like some sort of rubber sheet, deformed by the presence of mass. This is a conceptual drawing, not a rigorously calculated simulation.

It might not be obvious why we care about this, but there are situations in which we need to exchange information between observers in quite different spacetimes or measure things. Carrier pigeons and steel rulers won't do! With photons of light, the difference between an emitted signal and a received signal will tell you about the spacetime they are passing through. Changes to wavelength, travel time, the geometry of the path the signals take and other properties will inform you about the situation you're in.

Einstein described the relationship between how matter is distributed and the geometry of spacetime in his **field equations**. In theory, these would describe the behaviour of masses distributed in spacetime with whatever level of accuracy you desire. Anything from our bag of sugar, to a cloud of hydrogen gas, to a black hole would succumb to the power of mathematics. We could model orbits in the Solar System exactly and just as easily predict the dynamics of a multiple-star system in a distant galaxy. All we have to do is solve the field equations.

The problem is exact solutions are impossible except for a few so-called trivial cases (like a completely empty universe, for example).[12] We're forced to make

---

[12] You knew this was coming.

approximations even for situations you might think were simple, such as when two massive objects orbit one another.

I'm not sure I'd be brave enough to make some of these statements in a room full of theoretical physicists or mathematicians. 'Impossible' is certainly a word that could start several arguments. What I actually want to get across is that, more than a century after its conception, the General Theory of Relativity is still an integral part of our struggle to understand the nature of the Universe. Understanding black holes is an important part of that struggle, but only one part of it.

We usually refer to our advances as 'post-Newtonian approximations' because we have indeed moved beyond Isaac Newton's understanding of the Universe. Phenomenal amounts of research and hard work have gone into exploring these issues. It's important to remember that we have a fantastically impressive understanding about the workings of the Universe, it just isn't complete. And it isn't an all-or-nothing situation. Einstein's own approximate solutions to his field equations demonstrated why the **perihelion** of Mercury's orbit advanced as it had been seen to, something that could not be explained with Newtonian mechanics.

However, there are trivial cases. The easiest ones to imagine are vacuum solutions in which you don't actually have any matter present and spacetime is locally flat – that always seemed a little unsatisfactory to me – but there are others.

Of particular significance to us was an exact solution found by Karl Schwarzschild (1873–1916), who corresponded on the subject with Einstein in late 1915. It is a description of spacetime geometry outside of a spherical mass, assuming that mass has no electric charge and does not spin. There is quite an involved history unpicking the consequences of the mathematical solution, and modern relativity does not present it quite the way Schwarzschild did, but most effort was focused on what appeared to be two mathematical **singularities**: the first right at the centre of this spherical mass and the second at what is now referred to as the **Schwarzschild radius**.[13] Together, these are a mathematical description of an object (the spherical mass) from which light cannot escape – a black hole.

Inside this radius, spacetime curves round to the centre of the mass. *Anything* finding itself here is destined to end up at the singularity at the centre and that includes light.

Just to restate the above and dispel a common notion, you shouldn't think of a black hole as some black-coloured ball you

---

[13] The relevant papers were published in 1916, also the year of Schwarzschild's death. Astonishingly, he was doing this work (and had already made significant contributions in other fields) while serving in the army during the First World War.

could journey into and then head back out of when you've finished exploring. The warping of spacetime is such that inside the Schwarzschild radius, the only direction is towards the other singularity – the centre. There is no way to put the car into reverse! 'Backwards' is also towards the singularity.

An important extension to Schwarzschild's work was made by David Finkelstein in his 1958 paper 'Past-Future Asymmetry of the Gravitational Field of a Point Particle',[14] showing that the second singularity Schwarzschild identified could be considered as a one-way route to the first, regardless of your direction of travel. This 'perfect unidirectional membrane' can only be crossed in one direction: inwards. We refer to it as the **event horizon** and we'll be exploring this concept further later on.

---

[14] As far as theoretical physics papers go, this title is downright catchy.

We have determined that black holes can spin around and hold an electric charge. Both of these properties can affect the behaviour of particles or even spacetime itself near the black hole, but we'll only bring them up later on if we really need to.

Calculating the Schwarzschild radius

Putting equations into an explanation is a bit of a risk and I only do so at this stage because, for such a complicated subject, the Schwarzschild radius can be expressed in a relatively simple way.

$$r_s = \frac{2\,GM}{c^2}$$

Above: The Schwarzschild radius (r_s) is a function of what is known as the gravitational constant G (effectively a factor to ensure that the units of measurement stay meaningful), the mass of the object M, and the square of the speed of light designated c.

If we take some real-life examples, we see that the Schwarzschild radius for the mass of the Sun is only about 3 kilometres (see Figure 5). This means that if we, or nature, somehow compressed the Sun into a sphere about 6 kilometres across it would become a black hole. Another way to think about it is if someone was describing matter compressed to that extent, without using the phrase itself, you would know that what they were describing is by its nature 'a black hole'.

Realistically, several things would happen if you were to compress objects to such a degree, particularly something as massive as the Sun. Here, we're not imagining a situation where the Sun becomes made up of a type of material we might call 'black hole stuff'.[15] It's that if something satisfies the Schwarzschild

---

[15] For want of a better name …

Object	Mass (kg)	Schwarzschild radius (m)	... a bit like
Bag of sugar	1	1.48×10^{-27}	a 700-billionth of the radius of an atomic nucleus
Moon	7.348×10^{22}	0.0001	the width of a thick human hair
Earth	5.972×10^{24}	0.0089	the width of 1,100 red blood cells
Sun	1.989×10^{30}	2,950	ten *QE2* ocean liners end to end
Sagittarius A*	8.16×10^{36}	1.21×10^{10}	19,000 times the radius of the Earth

Figure 5. Some objects, their Schwarzschild radii calculated from approximate masses and some – possibly unhelpful – comparisons!

radius condition, the only thing it can be is a black hole.

This also explains why black holes fit comfortably in the domain of astronomy. We need large masses to create black holes. Try to work out the Schwarzschild radius for the bag of sugar you dropped earlier and it will be significantly smaller than the radius of an atom. Now, this isn't to say that small black holes aren't possible. Schwarzschild doesn't have anything to say about upper or lower limits. But to explore that further would involve delving deep into high-energy particle physics, beyond the scope of what we're considering here.

So, what actually is a black hole?

You've been very patient as I've set the scene and explained some tough theory. To put it simply: a black hole is a region of spacetime so curved by gravity that even light cannot escape it. We've imagined that such a thing as a black hole could exist, Einstein has set out the rules for a universe in which these situations might occur and Schwarzschild has presented the mathematical solution to one such situation.

It's all very promising, but a rigorous understanding of both rhino horns and horses doesn't exactly prove that unicorns exist.[16] Starting with all this theory is very useful though: by pushing some straightforward physics into extreme conditions, we convince ourselves that black holes are possible, even if we're just treating them like an interesting concept to begin with. Now we can explore our idea of a black hole in more detail, and then we can consider them in the context of 'real life'.

---

16 More's the pity.

The Anatomy of a Black Hole

Even if it seems strange, or a little counter-intuitive (believe me, the best is yet to come …), we have a fairly straightforward idea of what someone is talking about when they refer to a black hole. Of course, nothing in life is that simple, so let's make the whole thing more complicated!

Building up the layers

The very stuff of black holes is weird. Its structure is unsupportable even by the pressures inherent in normal atomic matter.

Mathematically, all the matter has collapsed into the singularity at the very heart of what we're calling the black hole. In fact, you can consider the black hole to be the singularity – a point of zero volume and infinite spacetime curvature. If the black hole isn't rotating, the singularity is in the form of a single point, and if it is, the singularity is a ring, still with zero volume and infinite curvature.[17] With mass squeezed into a single point, you can consider the singularity to be infinitely dense.

So, if we were building a black hole, the singularity is the centre and in some way it is the black hole. But what we really need to do is add in the extra layers that help us understand what is going on (see Figure 6).

---

[17] This might seem confusing, but a detailed explanation is beyond the scope of what we're considering here. If you want something really mind-bending, I urge you to consider a foray into the world of topology mathematics. You'll never look at doughnuts or helter-skelters the same way again!

Things heading to zero or infinity can be good fun for the mathematically inclined, but for everyone else it always feels a little ... uncomfortable at best. Worry not. There are some extra little things to consider right at the end, but by then we'll feel a lot more secure!

When people talk about black holes, they are often actually thinking not about the singularity, but the next layer out, known as the event horizon. Surrounding the singularity is that region of spacetime. Inside this nothing can prevent an infalling object moving further inward towards the singularity. Absolutely nothing. Spacetime, moving in any direction, curves round to the singularity inside the event horizon.

It is the point of no return or, more correctly, the boundary of no return. Its size? That's the value of the Schwarzschild radius. The result of crossing it is always the same: pass beyond this horizon and your destination is inevitable.

Messier 87 is a large elliptical galaxy approximately 55 million light years from Earth. A jet of atomic particles ejected by the dynamics of the supermassive black hole at the galaxy's heart is visible, but the black hole itself is not.
ESO

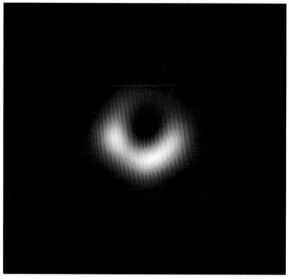

An image of the supermassive black hole in the galaxy M87, synthesised by the multiple radio telescopes of the Event Horizon Telescope Collaboration. The event horizon itself is a small central part of the dark region in the middle of this image and the black hole is about 6.5 billion times the mass of the Sun.
EHT Collaboration

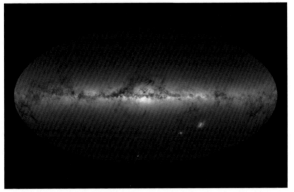

A view of the Milky Way as seen by the European Space Agency's *Gaia* mission. The Milky Way, our home galaxy, contains hundreds of billions of stars, but there are many unanswered questions about the population and distribution of black holes within it.
ESA/Gaia/DPAC, CC BY-SA 3.0 IGO

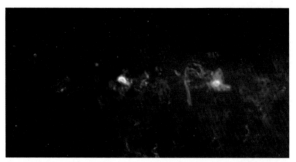

A radio, infrared, millimetre-wavelength composite image towards the galactic centre. The black hole Sagittarius A* is found here, though its nature means we gain our understanding of the object by its influence on the surrounding spacetime and nearby sources of light. We cannot see Sagittarius A* directly.
NRAO/AUI/NSF

There seems to be a relationship between the mass of a galaxy's central bulge and the mass of black holes within it, which suggests that their development or formation go hand in hand. However, although the dwarf galaxy ESO 495-21 is only about 3,000 light years across, it may contain a black hole of approximately 1 million solar masses. It seems we do not yet have the full picture regarding black holes!

ESA/Hubble, NASA; CC BY 4.0

The Crab Nebula is a beautiful example of a supernova remnant – the outer layers of a progenitor star that has exploded (an event known as a supernova). The light of the explosion reached us in 1054. Contemporary accounts state that it was visible during the day. However, while the outer layers were ejected outwards, the inner layers collapsed to form a rotating neutron star; despite the violence and energy of the supernova, conditions were not quite right to leave behind a black hole in this case.

NASA, ESA and Allison Loll/Jeff Hester (Arizona State University). Acknowledgement: Davide De Martin (ESA/Hubble)

This elliptical galaxy, Hercules A, is seen in a combination of visible and radio data. The pink lobes are invisible to the human eye, but show plasma driven out of the galaxy's centre by the supermassive black hole residing there.

NASA, ESA, S. Baum and C. O'Dea (RIT), R. Perley and W. Cotton (NRAO/AUI/NSF), and the Hubble Heritage Team (STScI/AURA)

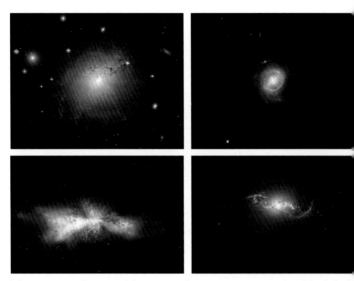

The centres of many galaxies are home to supermassive black hol[e]
some with masses billions of times that of the Sun. The preci[se]
relationship between the development of the galaxies and the bla[ck]
holes is not yet understood.

Clockwise from top left:

*NASA, ESA, NRAO and L. Frattare (STScI). Science Credit: X-ra[y]
NASA/CXC/IoA/A.Fabian et al.; Radio: NRAO/VLA/G. Tayl[er]
Optical: NASA, ESA, the Hubble Heritage (STScI/AURA)-ES[A]
Hubble Collaboration, and A. Fabian (Institute of Astronom[y]
University of Cambridge, UK)*
NASA, ESA and the Hubble SM4 ERO Team
NASA, ESA, W. Keel (University of Alabama, USA)
ESA/ATG medialab

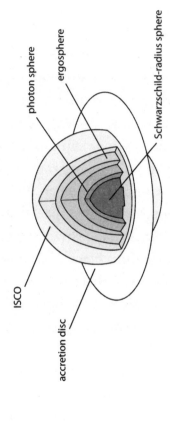

photon sphere

ergosphere

Schwarzschild-radius sphere

ISCO

accretion disc

Figure 6. A conceptual anatomy of a black hole.

41

In theory, it is possible for light to escape from the black hole if you haven't quite reached the event horizon. Stand just outside it and shine a laser directly away from the singularity (straight 'up', if you like). Those photons can escape. But while theoretically possible, it's probably not a situation you're going to find yourself in very often! The reality, anyway, is that photons of light nearby will still fall in, if they're moving at any sort of tangent.

Therefore, we come to another layer: the **photon horizon**, or photon sphere. At one and a half times the Schwarzschild radius for a non-rotating black hole, you could consider this the 'effective' size of the black hole you could see. No light from any source closer is actually going to be reaching a distant observer.

Between the event horizon and the outer boundary of the photon horizon is a region where light is pulled into unstable orbits (the exact nature of which depends on the angle at which you're emitting the photons and how close to the event horizon you are). The photons can't quite escape; they are going to spiral in eventually, but they may take a very long path to do so.

It is possible to have orbits that intercept the emitting object itself. The classic example is that photons emitted from the back of your head could orbit the black hole and hit you in the eyes, thus allowing you to see the back of your own head.

It's still not straightforward if you keep backing off. If the black hole is rotating, spacetime itself gets churned, a process known as frame-dragging or the **Lense–Thirring effect**. You've found yourself in the next layer, the **ergosphere**. The precise character of this region depends on the angular momentum of the individual black hole, so it isn't a straightforward thickness or even shape to define. But the behaviour of objects within the ergosphere is the fascinating part.

Here it's as if you're being sucked down the plughole in a draining sink. You cannot remain stationary unless you power yourself in the opposite direction. Close enough to the black hole, your opposing motion would have to be greater than the speed of light just to stay still, and since this is impossible, you fall inwards.

So far, things aren't very positive if you're trying to avoid the black hole. Beyond the event horizon, the only direction that can be said to exist is inwards. Beyond the photon sphere boundary, there is some wiggle room about what your precise trajectory is going to be, but your eventual fate is still to spiral in. And for spinning black holes, you have to exert increasing amounts of energy to stay still or, you guessed it, you'll spiral inwards.

Keep backing away and you will eventually find yourself in flatter, calmer spacetime, except for a last little hurdle. Newtonian mechanics might suggest that you can put yourself into a nice stable orbit at any distance, but its mathematics are only approximations when we consider the extreme environment near a black hole. Relativity has other, more accurate, ideas, but they're not always encouraging. For any object (whether atomic particle, spaceship or that bag of sugar we were playing with earlier), there is the **Innermost Stable Circular Orbit (ISCO)**. If the black hole is spinning, things get even more complicated, but in this scenario there is no spin and the ISCO is at a distance of about three times the Schwarzschild radius. Get any closer than the ISCO and you can no longer stay in a stable orbit, which would otherwise have prevented you from moving yet closer to the black hole. You'll then be trapped in a spiral towards all the other problem-causing layers.

Further back, things become more understandable. Theoretically, we can get into a stable orbit around the black hole and stay there without spiralling inwards. Better yet, and probably more sensible, we can fly away from the whole mess! Of course, a 'realistic' situation in which you're trying to explore anywhere nearby with a spaceship would have all the real-life complications of building something to withstand the gravitational pull, using fuel to fly around and that sort of thing. Probably best if we continue to explore using our thought experiments, at least for now.

Peeling off the layers (or 'Help, I'm falling into a black hole!')

One of the most common questions astronomers are asked is: what happens if I fall into a black hole? Quite why it seems of genuine concern to so many people is beyond the scope of this project, but considering it seriously is a rather good way of putting the previous section into context and making sense of the terms we've just defined.

Let's say you've been exploring the galaxy and got a bit too close to a Schwarzschild black hole without realising. I, eminently more sensible, have stayed far away on a distant space station, but I'm willing to help if I can! And, being generous as well as sensible, I gave you an amazing spaceship that can do almost anything. The question is, will almost be enough?

In our situation 'too close' means you can't put yourself into a stable orbit, so you're going to have to expend energy to maintain an orbit. You've crossed the ISCO. No problem, your spaceship has powerful and efficient engines. (Thankfully, the black hole isn't rotating, so ergosphere effects aren't something to worry about.)

Alas, other stuff is drifting towards the black hole too. Molecules of gas, a bit of dust, perhaps some other unfortunates without a spaceship as good as yours.

47

You've found yourself making up part of the **accretion disc** of infalling material and it isn't quite as collegiate as it sounds. Though not part of the black hole itself, this material heats up due to friction and produces electromagnetic radiation, and it falls further inwards. The wavelength and intensity of the radiation depends on the mass of what you're spiralling into – the accretion discs of protostars, for example, radiate in the infrared, which might not be too bad because that emission is quite low-energy. But in the case of black holes, the gravity and friction are sufficient to blast you with high-energy x-rays or gamma rays. Sorry!

Because the disc is flat, moving around what you're going to think of as the equator of the black hole, you might try to get yourself 'above' or 'below' the disc, but you'll be pulled back into alignment.[18] You haven't even reached the weird stuff – this is fairly conventional mechanics – though already you might realise you're in trouble.

It's probably time to start signalling for help so you send a radio signal to the distant space station from where I'm observing things as best I can. Realistically, what can I do? As you get closer to the black hole, the signals you're sending are undergoing increased **gravitational redshift** as the photons climb out of the gravitational well. The space station is literally receiving different wavelengths

---

[18] Bizarrely, I would actually be able to see regions behind the black hole from my position thanks to accreting matter. Some of the radiation it produces doesn't escape directly away nor fall into the black hole but heads off at a tangent and is curved by gravity into my line of sight. In theory, I would see a sort of increasingly compressed rainbow of images of the accretion disc behind the black hole.

than you're transmitting, but while I'm busy retuning the receivers, you're becoming ever more worried about the tidal forces acting on the spaceship.

The gravitational gradient as you get closer to the black hole is increasingly steep, so the difference between the gravitational pull on the parts of the spacecraft nearer the black hole and those further away grows. You're beginning to stretch, resisting that stretch by the strength of the material making up your spacecraft. But in extreme situations the tidal forces will pull you apart, a process known as **spaghettification** (see Figure 7).

Spaghettification is not inevitable. Black holes of different masses will have different gradients, so with supermassive black holes it is perfectly possible to pass the event horizon with no ill effects. Again, this is not to say that the gravitational pull isn't strong, just that the gradient isn't too extreme. Let's assume this is the case.

Figure 7. Spaghettification! The curved line represents a section of the surface of the black hole. In the first panel, the astronaut's height and width correspond as expected. As he moves closer to the centre of the black hole, he experiences slight compression horizontally and elongation vertically. In the last panel he is closer still and the compression and elongation of his form are even more dramatic.

Unfortunately, other stuff seems to be falling into our black hole too. Although a bit of company might seem welcome,[19] infalling particles spiral into the black hole in a turbulent flow, rubbing up against each other. As we've seen, the accretion

---

[19] You've been signalling for help – what's taking so long?!

disc formed emits radiation due to this friction, and because of the immensity of the gravitational pull particles are accelerated up to significant fractions of the speed of light. The result is highly energetic radiation, including things like powerful x-rays. The black hole may even produce tightly focused astrophysical jets of ionised matter (sufficiently powerful and with velocities high enough to be referred to as relativistic jets).

Jet beams may extend millions of light years, so hopefully we would have noticed them before now.[20] They're also complicated things with unanswered questions surrounding them, so we won't dwell on them too much. Significantly, they align with the axis of rotation, whereas we are approaching almost perpendicular to it in the accretion disc.

---

[20] Although maybe not if you've started falling into it!

You have enough to worry about, so let's assume the jets aren't a problem.

In fact, suppose you're falling into an otherwise entirely quiet black hole and there are no jets or an accretion disc. Now the real fun begins. By the way, I'm happy enough to make all of these assumptions, but it is worth noting how many we're making just so the 'ordinary' mechanics of a black hole don't destroy us!

The gravity of the black hole warps spacetime itself. In our case we're interested in a process known as **time dilation,** another relativistic effect. It is a particularly complicated issue, so we'll treat it with a light touch here. But even so, if we want to describe its effects, we have to be careful about our point of view. Things will look quite different to observers in different positions.

Let's assume I am still in that distant space station you signalled for help. Time dilation means that from my perspective you actually start to slow down as you fall into the black hole. As far as I'm concerned, time is literally passing more slowly for you than for me.[21] By my calculations, your passage through time is going to crawl to a halt as you approach the event horizon. Remember, I'm finding it increasingly hard to even detect you since your radio signals for help, and any other photons you're emitting, are being stretched to longer and longer wavelengths. You, meanwhile, are experiencing time passing normally.

Hold on, isn't time ... time? How is that clock working differently? Well, time itself flows at different rates depending on what is happening. These aren't figures of speech: objects move in spacetime at different rates (not just in the normal three dimensions associated with their velocity as we normally understand it, but in time too).

---

[21] It's not a trick – if I could see the hands on your watch, they would be ticking more slowly.

We're really noticing the effects because of the extreme situation we've engineered, but to a greater or lesser extent these distortions are happening wherever you are.[22] If everything else was fine, you'd simply pass the event horizon and keep going towards the singularity. Of course, the words 'fine' and 'simply' are doing a lot of work here!

What's happening 'outside' the black hole while that's going on? Well, sadly I've had to give up on being able to reach you or even get a signal from you. You've passed the point of no return. Sorry! From your point of view, there is no outside to reach any more, since every direction leads you towards the centre of the black hole. From my point of view and to all intents and purposes, you've ceased to exist.

This is why black holes are just ... the end, both in space and time. Effectively,

---

[22] Good luck trying not to think about this when attempting to get to sleep tonight!

things crossing the event horizon play no further part in the Universe. They cannot signal anyone and vice versa. And when we say signal, we don't really mean sending a coherent message. We simply mean transmitting any information at all.

Imagine you are an astronomer sitting in a laboratory beside an astonishingly sensitive detector that somehow reacts to a single, solitary photon of any wavelength emitted from beyond the event horizon of a black hole. When it does, it will set off an ear-splitting siren, so you definitely won't fail to notice. But day after day, year after year, there will be nothing but silence. As far as that detector is concerned, there is simply nothing there. A complete information void.

In summary: don't go near black holes!

A brief aside: if that last part seemed like a rush of information, don't worry. There is a lot to unpick, and some of the issues get quite philosophical. What does it mean to have regions of spacetime that don't seem to participate in the rest of the history of the Universe?

Further, when we've talked about that 'unidirectional membrane' (see page 33), it's all too easy to picture a weird sort of black balloon where travelling from the event horizon to the centre of the black hole is like passing through the balloon skin (or maybe that's just me). Now obviously there is no balloon, but described one way it really does seem like there is at least a ... layer, let's say. A structure in some sense of the word. But a different (and arguably more accurate) way to consider the event horizon is the skin of 'events' that occur just before they no longer contribute to the Universe. It is a series of frozen occurrences. Weird, right?

The situation is vastly different depending on the observer's position, so it is hard to build a coherent description of what is actually going on and this extends to purely mathematical treatments too. Plus, we'll see later on there are potential complications and additional considerations!

There is no point pretending we've solved these issues. They're a matter of ongoing research and discussion and doubtless a good starting point for some interesting science-fiction stories and thought experiments. Imagine, for example, how you would describe my heroic rescue attempt of your spacecraft just before it crossed the event horizon! Those still on the space station would see an agonising, increasingly drawn-out approach as I started to 'slow', but would I be approaching relatively fast enough to reach you and help?

On a slightly more serious note, we're going back to those 'thought experiments' as a useful way to examine the ideas surrounding black holes. Trying to imagine the experiences of people in different specific scenarios is quite an effective way of building up an overall picture of the situation.

Creating a Black Hole

We've examined quite a bit of black hole theory in a short space of time. But are they actually out there? Randomly searching doesn't seem particularly sensible, so let's examine situations that might produce them so we know where to look.

Very simple and fantastically complicated

Stars are essentially a battle between the collapse of matter under gravity and the outward pressure powered by nuclear fusion processes. However, the battle can have different outcomes, which can lead us to black holes.

Fundamentally, atoms don't like being compressed too much. They have a lot of leeway, but they aren't just free to squish ever closer together. In fact, they resist. **Electron degeneracy pressure** builds up – a quantum-mechanical consequence that fights further compression. We're not talking about when you try to cram too much into your suitcase, but about much more extreme circumstances that place us firmly in the realm of astrophysics.

Electron degeneracy pressure is most commonly brought up when discussing the fate of low-mass stars like the Sun. At the end of its life, our star will push off its outer layers, eventually creating a lovely planetary nebula. But the interior will collapse and without the outward nuclear-fusion-powered pressures there will be enough mass to create a white dwarf.

The white dwarf is electron-degenerate matter. It's incredibly dense. No more nuclear fusion, but plenty of thermal energy left to emit. The dying ember of a star, if you like. It will cool and has the potential to create a black dwarf star.[23] However, if there is even more mass, not even electron degeneracy pressure is enough to prevent further collapse. This upper limit, the **Chandrasekhar limit**, is a little less than one and a half times the mass of the Sun (though the precise limit for a specific case would depend on other conditions including how the material was spinning), and frankly that's nothing. There are plenty of stars with much higher masses. What then?

Well, atoms have another line of defence. As they continue to be compressed, electrons and protons actually combine to create neutrons. Neutrons don't like being

---

[23] But that process takes so long our Universe hasn't actually created one yet.

compressed too much either and **neutron degeneracy pressure** builds, resisting further collapse. This can result in neutron stars – even denser, even stranger neutron-degenerate matter.

To demonstrate, let's throw some rough numbers around (there is enough complexity to degenerate-matter physics to write many books about the subject, so we'll keep things quite approximate). A neutron star with the mass of the Chandrasekhar limit would only be about 20 kilometres wide. As you'll recall, we're not too far off black hole sizes so you might guess, correctly, that we're in a pretty extreme environment. Standing on the neutron star's surface, you'd experience 100 billion times the gravitational pull you would standing on Earth. It would be comparable to a chicken egg 'weighing' somewhere between 5 and 10 million tonnes (remember we're being rough with the numbers and not worrying too much about the difference between mass and weight).

Neutron stars have their own story to tell, but what concerns us is that they too have a limit, known as the **Tolman–Oppenheimer–Volkoff limit**. This is a little over two solar masses (that is to say, twice the mass of the Sun) and not even neutron degenerate matter can resist the collapse to become a singularity. We have created a black hole.

Admittedly, we've just glossed over a lot of quantum mechanics and atomic theory, so let's look at it in what I think is its most straightforward form: gravity attempts to pull things ever closer together, but matter resists this compression. However, when there is enough matter, the atomic structure itself cannot withstand the force of gravity generated. It is a self-defeating situation. Gravity wins. Collapse is inevitable and with it comes the creation of a black hole.

In one sense then, black holes are among the simplest things in the Universe to create. Stuff plus gravity plus time equals a black hole.

Now, a couple of solar masses isn't that much really, at least not in the grand scheme of things. With billions of stars in the galaxy, and billions of galaxies, shouldn't the Universe be absolutely littered with black holes?

Supernovae

In those battles between gravity and outward pressure mentioned above, it is possible for stars far more massive than allowed by the Tolman–Oppenheimer–Volkoff limit to exist.[24] Gravity is a rather weak force to be honest, at least relative to nuclear and electromagnetic forces, so

----

[24] At the time of writing, estimates for the mass of a star known as R136a put it over 300 solar masses.

enormously massive **progenitor stars** can keep themselves stable.

But not for ever. Eventually, stars cannot fuse material in their cores and gravitational collapse takes over. In the end phases of a star's life, this can produce the catastrophically violent ejection of its outer layer and an enormous, transient increase in luminosity.

An explosion, basically – what we call a **supernova**. For a brief period of time, it can outshine its host galaxy as the outer layers are thrown off at thousands of kilometres per second. Meanwhile, no longer supported by nuclear fusion, the core collapses under gravity and this is where we can satisfy the conditions to create a black hole.

This is quite a glib explanation of what is a complicated set of conditions and processes. There are different types of supernovae, and a number of physical mechanisms involved, but I'd need another book (or two!) to cover those. They're also rather rare, meaning direct detections are also uncommon. My favourite historical example is when the light of a supernova reached us in ad 1054. How do we know? Well, people all over the world wrote about what appeared to be a 'new star' appearing in the sky. At a time when international scientific collaborations weren't exactly standard practice, there are records of how this phenomenon (which no one back then would have understood) could be seen during the day before fading away.

In its place, we now have the famous Crab Nebula, created by the escaping outer layers of the progenitor star. And at its heart is the Crab Pulsar: a rotating neutron star, which we can now 'see' thanks to modern technology. Not quite massive enough to have created a black hole though!

A progenitor star of at least eight solar masses is needed to create a supernova.

Less massive stars will either push off enough surface material or be able to rely on the electron or neutron degeneracy to keep themselves stable as white dwarfs or neutron stars, as described previously.

Figuring out how common these occurrences are, the population statistics of supernova-creating stars and other matters are active areas of research, and it goes to show that lots of astronomical issues are tied together. Parts of the black hole puzzle can be chipped away by nuclear astrophysics advances and vice versa. It's hard to tell where the next illuminating insight will come from,[25] but it can be unexpected.

To highlight just one project of interest, the Rubin Observatory currently under construction in Chile has an astonishingly high-resolution camera, a wide field of view and can thus cover large swathes of the sky in detail rather quickly. Perfect,

---

[25] See what I did there!

67

we hope, for seeing transient events like supernovae! This will undoubtedly help improve estimates about stellar populations and event rates. In fact, one of the principal issues concerning the project is how to process the vast quantities of data the observatory will collect each night. While we often remember the astronomers working on theory and the teams responsible for creating the amazing instruments used to observe the Universe, there are also huge efforts to actually analyse the collected data and produce meaningful information that can lead to further insights.

Mergers and accretion

You don't always need to begin with something enormously massive to create a supernova. We could exceed our mass limits by merging multiple less massive objects – by 'feeding' a white dwarf material from a partner star in a binary system, for example. In theory, you could feed in anything that gets close enough for gravity to pull it in. Two colliding neutron stars would be effective too. Individually we might think of them as 'not quite there' in being able to create a black hole, but the combined masses would tip the scales.

Of course, space is extremely big and mainly empty. Still, there are enough mergers going on that we might expect a large number of black holes. Finding them is the difficult part (which we will tackle later), but it means our estimates of the numbers of black holes and the distributions of their sizes are rather poor. It is hard to be confident about the statistics of a population with such a small sample set. This should improve with further study, although it's difficult to know exactly how things will change. Of course, new generations of telescopes may bring about fantastic leaps in our understanding, but even with current efforts we will continue to build up a more detailed picture as time passes.

A very common question is: can black holes eat other black holes? The simple answer is: yes, they can. Again, black holes are simple things in some senses – don't worry about the mathematics, just try to remember what black holes are. They're regions of spacetime where the gravity is so strong that nothing can escape. Two black holes can collide but their nature means that rather than bouncing off each other like billiard balls they will merge together. To form something even stranger? No, simply a more massive black hole.

Galactic centres and some problematic unknowns

We have a rather massive black hole, a little over 4 million solar masses, in the centre of our galaxy called Sagittarius A* ('Sagittarius A-star'). We classify it as a **supermassive** black hole, as opposed to the black holes we would see produced by supernovae ('**stellar-mass** black holes', which may have masses of a few solar masses, perhaps as many as a few hundred). How did Sagittarius A* get so big and is this usual?

The truth is, easy classifications aren't always possible in an active research field. Confirmed detections of black holes in galactic centres have been made in only a handful of cases, but they suggest a correlation between velocities of stars in the central regions of those galaxies and the mass of the black holes. If this relationship holds true, the dynamics of stars in other galaxies suggest that supermassive black holes are common. They exist in almost all massive galaxies and some have billions of solar masses!

Can this really be the case, and if so, why? Again, research is ongoing. It certainly seems that there is a relationship between galaxies and black holes, but the precise reason escapes us for the moment. It could be that the dynamics of material in galaxies 'feed' stellar-mass black holes until the accumulated mass is in the supermassive range (on which opinion as to the parameters differs). Alternatively, is there some system by which black holes formed in the early Universe anchor enough material to form massive galaxies and things develop further from there?

What about something that is neither stellar-mass nor supermassive? Well, in theory there should be things we could classify as 'intermediate-mass black holes'. Confirming the existence of these objects – too massive to form from a single star's collapse, but not supermassive – could shed light on the whole subject. Currently we have a few candidate detections, some of which are extremely promising. It may be that we've spotted that missing piece of our jigsaw puzzle. Astronomers will doubtless argue about this for a while and be itching to add other categories.[26]

It seems certain that there is more at play here than what we've discussed. A more complete picture would probably be very concerned with early Universe dynamics and by extension things like dark energy and dark matter. We're beginning to stray somewhat, so let's investigate how we got this information in the first place.

---

[26] I hereby predict that someone will start using the term low/high intermediate-mass black hole, or some such thing at some point.

Finding Black Holes

We have a theoretical idea about our prey – we can describe a black hole well enough and the places in which one might be found – but how do we go from there to spotting one? On the face of it, this isn't an easy thing to do. After all, we're looking for something that literally won't let light escape: 'seeing' in the conventional sense doesn't seem a particularly useful way forward. What now?

Looking for the invisible

Think about what happened when you fell into that black hole earlier. There was plenty of stuff going on before you were eventually consumed. Might it be possible to use infalling matter to 'map out' the region surrounding the singularity, perhaps even right up to the event horizon itself? Yes, this is indeed a way to go about black hole detection. John Michell was right!

Let's return to Sagittarius A* in the heart of our own galaxy, the Milky Way. Known since the 1950s as a radio-bright source (that is, it was very luminous when astronomers pointed radio telescopes at it), it was thought of as a pretty exciting region to research but it wasn't exactly clear what was going on. In 2008, a decade-long initial collection of near-infrared data revealed that a star (given the name S2) was orbiting this region at high velocities. S2 is more massive than the Sun, perhaps as much as 15 times as massive, so whatever was pulling it into orbit must have been massive indeed. An original estimate of almost 3 million solar masses was made for this mystery object. Further, the mystery object itself could not be seen. So the question was really: what in nature can have millions of solar masses and not be luminous? The best answer was a black hole.

Subsequently, other nearby stars, also being pulled into orbits by the enormous gravity of Sagittarius A*, were tracked. After further research, a figure of approximately 4.31 million solar masses was reached and although different analysis techniques have not agreed exactly, they are in broad agreement. This is considered the best empirical evidence that there is indeed a supermassive black hole at the centre of the Milky Way.

There was more good news to come. Sagittarius A* is fairly quiet, but not silent. On more than one occasion the black hole appears to have fed on something, causing the accretion disc to brighten. Additionally, some of the nearby stars are moving incredibly quickly. At the time of writing, the star S62 is travelling at about 7% of the speed of light and is at a distance of a little more than 200 times the Schwarzschild radius (of Sagittarius A*) from the black hole. This is close enough and fast enough to show that general-relativistic mechanics fit the observations better than Newtonian mechanics, lending even more weight to Einstein's ideas.

Truly invisible?

There is a possibility that black holes do actually radiate as a result of quantum-mechanical effects through a process explored most famously by Stephen Hawking. Currently, relativity and quantum mechanics have yet to be combined into a universal theory that covers every situation. It remains unclear whether this is even possible, so we're certainly in an interesting area of study. Still, truly vast amounts of progress have been made in both fields and there is definitely the will to create a unified framework. Given the extreme situations that black holes naturally throw up, it is perhaps no wonder that they are great laboratories to explore how we might do this. Nevertheless, we don't have time to delve too deeply into the quantum realm, so we'll just extract a few of the ideas we need to illuminate the bigger picture.

One interpretation strongly suggests that quantum fluctuations create particle–antiparticle pairs of virtual particles, which can combine and effectively cancel each other out: literally the appearance and disappearance of something(s) from nothing. At the macro scale, which we inhabit in everyday life, this seems quite bizarre and without cause and indeed you might think of these pairs as some fundamental expression of the Universe's inherent uncertainty (in the correct scientific sense, rather than just meaning 'weird', though it certainly is that).

In essence, the Universe doesn't notice these fluctuations as such, since the average total energy remains at zero over anything but the tiniest timescale. Creation and destruction cancel each other out. But if a particle–antiparticle pair was created very close to the event horizon, one part could be 'eaten' by the black hole, just like anything else, while the other managed to escape being swallowed.

Since the total energy seems unbalanced by this process (another interpretation of the same effect is that the black hole 'boosts' one of the virtual particles into becoming 'real'), we now have an issue. We can deal with it by assuming the 'eaten' particle has negative energy. In response, the black hole itself loses a tiny bit of mass and the balance is restored.[27]

The result is that black holes 'evaporate' by the emission of what is known as

---

[27] It's an elegant and fascinating idea, but, between you and me, I'm not sure I grasped anything but the basic principle until someone explained it for a third time.

'Hawking radiation', which is theoretically measurable. However, what we refer to as the temperature of a black hole is inversely proportional to its mass: really small ones 'boil off' quickly, and big ones absorb more cosmic radiation than they are releasing. The balance point is a tiny black hole roughly the mass of the Moon, so greatly less massive than something typically produced by supernovae, for example.

In fact, at the moment this may be a completely moot point. The amount of radiation produced is so small that it is undetectable by any current experiment or even anything we're dreaming up, as far as I'm aware. We may eventually be able to 'see' black holes directly, but for now they remain invisible.

Creating a picture of a black hole

In April 2019, a global network of telescopes released an image of a black hole in the centre of an elliptical galaxy known as Messier 87. Naturally, this made headlines around the world, but it is sensible to examine the issue a little more carefully. The image created was not, as you might have guessed, like taking a photograph. Instead, it was constructed from the data collected by a telescope network, effectively working together as the Event Horizon Telescope (EHT). It was as much a triumph of data processing and analysis as it was an exploration of relativity.[28]

The combined data, once processed, has thus far produced a singular image, which appears consistent with the predicted look of a rotating black hole. Once again, the actual central region is not luminous (and remember the event horizon boundary

---

[28] Honestly, it's worth looking into the work of the EHT network. Apart from anything else, it involves physically flying hard drives on planes around the world to collect the quantity of data required.

would be smaller still), but we see it ringed by an accretion disc.

Listening for the invisible

An alternative detection process is to 'listen' for gravitational waves emitted by mergers involving black holes. A more thorough description of gravitational waves is beyond the scope of this book, but, put simply, moving asymmetrical masses cause distortions to spacetime itself, which head out at the speed of light. They're a prediction of relativity, and it is sometimes helpful to think of them as a bit like soundwaves, or using the phrase 'ripples in spacetime', though neither of these analogies really holds up. Additionally, although they have wavelengths and polarisations, gravitational waves are not any sort of electromagnetic radiation. They are very much their own thing. You can't 'see' them in the same way we see x-rays or microwaves, or any other part of the electromagnetic spectrum.

The asymmetry is important, especially considering a black hole would be pretty much a perfect sphere. But were two black holes to collide, or were a black hole to merge with something hefty like a neutron star, we'd be in business. Gravitational waves sit firmly in the field of astronomy for quite a simple reason. They are incredibly weak. Far too weak to create artificially on Earth (and there are other difficulties too, it's a complicated issue). We rely on the immense masses of things like black hole collisions, and even then it is next to impossible. But crucially, not impossible! In fact, despite Einstein predicting gravitational waves in his General Theory of Relativity in 1915, the first confirmed detection event was made by the Laser Interferometer Gravitational-wave Observatory (LIGO) facilities in

September 2015 (announced publicly in February 2016), so it only took about a century.

GW150914, as the signal was called, was a merger of two black holes followed by the 'ringdown' of the resulting larger black hole (much like the fading note of a struck bell, the black hole created a burst of gravitational waves, the frequency and strength of which ebbed, until it settled completely). The energy carried away as they merged was roughly the equivalent of three solar masses being annihilated; in one sense the event was 'brighter' than all of the stars in the observable Universe. But again, we don't see them electromagnetically. In fact, the detection is made by measuring the position of mirrors using lasers as the spacetime distortions pass through the detector. The gravitational waves shifted the relative positions by somewhere in the order of 1-millionth the radius of an atom. Gravitational wave detection is difficult!

Subsequent detections using gravitational waves have provided further insight into black holes – in the first instance, they confirmed that stellar-mass black holes definitely exist and that such mergers actually happen. Having outlined what seems to be a sound theoretical understanding and followed that up with multiple detections, it might be sensible to question how any of this helps. What does this really teach us? These are the kinds of questions scientists hate to answer. Despite the long delay getting to this point, we really are just at the start of the era of gravitational wave detections. Over time our catalogue of events will build, providing more accurate statistical models of black hole populations, better measurements of their merging parameters and possible (as yet undreamed of) insights into the objects themselves and their mechanics. Watch this space!

How many black holes are there?

This is a much easier question to ask than to answer. The easiest and most accurate way to summarise would be to say, 'we're not sure'. This is a common issue in astronomy. Each new detection of a black hole helps build up an estimate of the population, but we have the problem of trying to consider the whole Universe at the same time. A really detailed population study of one galaxy, even if you had one, might not be particularly informative about the hundreds of billions of other galaxies.

Announcements seem to be made every day. There are new gravitational wave detections, more accurate measurements of supernovae rates, better surveys of the galaxies (and attendant stellar population studies), not to mention the advances in nuclear astrophysics, quantum mechanics ... the list goes on. Each small advance reveals a bit more of a picture still unknown in size and complexity.

In fact, it's hard to say anything definite at all. We have discovered stellar-mass black holes and supermassive ones, but evidence for intermediate-mass black holes is very thin on the ground. Many (but not all) galaxies have central black holes, but a proper idea of what's going on in the nearby galactic neighbourhood is lacking at the moment. If they're not active in some sense (colliding, feeding, etc.), black holes are extremely difficult to detect. Enlightening ourselves in this field will be slow, careful work.

In May 2020, astronomers announced that they had discovered a black hole within 1,000 light years of Earth. There may even be others, yet to be detected, that are closer. Perhaps many! Anyone who is worried about falling into a black hole might feel a twinge of panic at this news (which is just one story related to black holes to have emerged of late). Quite often there is the notion that black holes are 'on their way' to the Solar System, or they're personified as hunting us down so they can gobble us up. To combat such ideas, I can only remind you that black holes are physical objects governed by rules, even if we don't completely understand them. But they are natural things, like stars or planets, and not many people worry about those being out to get us. Rest assured, a rogue star careening through the Solar System would probably ruin our day just as much. If nothing else, I'm pretty confident that there is no black hole in the local region so close that we'd have to worry about it in our lifetimes, even if it was hurtling towards us at close to the speed of light.[29]

---

[29] Pretty confident ...

What Does This Teach Us?

To talk about black holes, we've considered subjects ranging from stellar evolution to quantum mechanics, but only touched on each one very lightly. Literally lifetimes of study are waiting for those who want to know more. Rather than examine something specific in more depth, let's recap a little and examine a few issues emerging from the understanding we've already developed.

A quick look back

We ought to remember that black holes started as an idea: a consequence of what should happen if the Universe worked the way we thought it did. The idea itself, albeit modified and greatly improved over time, survived various changes to our universal framework. Actual detections came later, but the idea could not be shaken off. One way or the other we were going to have to go hunting for these things.

Why? Because they were important laboratories to test our framework, perhaps to destruction; incredibly massive, gravitationally powerful and producing situations wherein even the speed of light would not be sufficient to counteract the influence of the black hole. And if they didn't exist? We'd be in trouble because our framework seems to cover every other situation almost perfectly.

At this point, you could argue we're interested in the fine details because we know a lot. We can describe a great deal, if not all, of black hole dynamics mathematically. We even have a picture of one. In some sense, you could say that we're thinking philosophically about what they mean.

Some other considerations

If you were so inclined, you could now turn your attention to black hole thermodynamics, the Blandford–Znajek process, or maybe even exotic-star black hole alternatives. And of course, there's enough mathematics to boggle the mind! Unfortunately, we don't have the space for everything, so I'll draw your attention to just one more oddity if you're looking for somewhere to begin further study.

The oddity starts like this: black holes themselves are quite simple to describe mathematically, even if the dynamics of the surrounding spacetime (or, crucially, infalling particles) are complex. All you need regarding a black hole is its mass, angular momentum (you're considering its spin) and charge. This seems great at first.

Simplicity would be quite welcome after everything we've been trying to think about.

Remember that we're considering the black hole in and of itself. The process of describing a black hole for the purposes of finding it would include things like its position in the sky and distance and the corresponding orientation with respect to your detector. All of these would be time-dependent too, to say nothing of the additional parameters required to describe a binary black hole system or something more complex. But these are not the intrinsic parameters of the black holes.

But what happens to the information regarding the matter used to form the black hole? Uh-oh. Quantum mechanics isn't happy with the idea that this information is just lost, and there is debate about whether this is simply a problem for quantum mechanics itself, or if it spills into a situation that drags essentially every field of physics into the argument.

When black holes seemed like everlasting objects, the idea being that they could trap the information so it was inaccessible for ever but not actually destroyed, was one workaround. If black holes evaporate over time, however, the problem rears its head again. It's the so-called black hole information paradox.

One potential solution could be contained within the mechanism causing the problem (which has a pleasing symmetry, at least). The quantum information could be entangled within the Hawking radiation emitted by the black hole. In effect, the information returns to the Universe as the black hole evaporates. So if you fall into a black hole, the idea is that you do indeed return but in an information-theory sort of sense. You, as a person, don't pop back out unharmed (sorry). You were a goner as soon as you started getting pulled in. Once again: black holes, fascinating, but best avoided.

Even this might not work as a solution. If, eventually, all the information has to be returned, how is this tracked? Are radiated particles quantum-entangled somehow with both the black hole and previously emitted particles? Could you say that the very final particle emitted is entangled with everything emitted before it, but no longer the black hole (which has ceased to exist, because it just evaporated fully)? This is, as yet, a mystery and one of the unsolved problems of physics.

Wormholes, teleporters and time machines. Oh my!

We've firmly established that things get weird near a black hole, but also that our current understanding means we probably don't know entirely how weird. A complete mathematical treatment of black holes in all circumstances doesn't exist, unfortunately. Nevertheless, what we currently know suggests some intriguing possibilities.[30] These are quite often fodder for science fiction and thereafter endlessly squabbled about when discussing their realism (or lack thereof in the interest of a satisfying narrative).

First, the extreme time dilation caused by gravity near a black hole would make for an (admittedly rubbish) time machine of sorts. If you were to explore the region close to the black hole without getting so close that you plummeted into the singularity, you could return to find that years or centuries had passed from my frame of reference, while your own frame of reference suggests that it was mere moments. You have effectively travelled to the future. In fact, exploring close to the black hole means this sort of issue is unavoidable and, really, the only question is to what extent will it occur.

---

[30] If you can figure out how to do or survive anything in the remainder of this chapter, tell someone, for goodness' sake. Science prizes and vast wealth surely await.

Rather than stepping through a portal or getting inside a fantastical machine, we're putting the travel into time travel. Journeying near a black hole allows you to leave what you think of as the 'normal' passage of time, returning to a spacetime that has rushed on at some enormous rate without you experiencing all that time pass.

Let's make things stranger by imagining that we've created or formed a black hole in the shape of a bagel. Topologically, this is actually still spherical, so we haven't broken the rules yet. Now let's cut it in half to create two, still bagel-shaped, halves.[31]

To prevent their collapse into a sphere, we maintain the bagel shape using some sort of (for the moment) exotic matter with negative energy density. This is mathematically valid, even if we don't know of anything that would behave in this way, or whether its

---

[31] You'll be on Nobel Prize number two by this point. At least.

existence is physically possible. Assuming we can do this, the spacetime of these bagels is in theory linked such that a journey through the hole in one half of the bagel takes you out through the hole of the other half, regardless of their separation.

It seems as though we have an odd sort of teleporter, providing we can move the two halves apart. We might move these a significant distance from each other, but the joined spacetime means that an explorer on the right trajectory goes into one black hole and emerges from the other almost instantaneously.[32]

---

[32] Astrophysicist Kip Thorne (2017 Nobel Prize winner for his work on LIGO) has done work on this subject looking at what are referred to as Einstein–Rosen bridges (rather than wormholes), linked spacetimes, networks of black hole teleporters and a great deal more. Some of his ideas have made their way into science-fiction movies and popular fiction – they're great fun to explore.

Things are already rather strange, but they can always get weirder. Now imagine that we could move one half of the bagel black hole very close to the speed of light. Let's call this half B, and the stationary half A. Again, time dilation takes effect, and B experiences only a very short period of time passing. We have a link between two objects and can travel between the two in an instant, but for the spacetime in B almost no time has passed. Jump from B into A and we emerge not in A's timeline, but in B's. From A's point of view, we've jumped into the past.

Don't worry if your head hurts, it's worth reading the previous bit more than once! What we've constructed is either an extra-strange teleporter or a time machine of sorts. Unfortunately, it is a bit limited and not just because it is difficult to imagine how you would construct it in the first place. What we're suggesting is that black holes can make time machines that can take you back to the moment when you made the time machine, but no further. I'm afraid to say it isn't like the typical science-fiction idea. You can't jump into a contraption and watch the fall of the Roman Empire. The Romans didn't have a time machine, so that part of the past will remain inaccessible. We can only go back as far as the first successful creation.

There are two compelling ideas that arise from this constraint. The first is that we haven't uncovered secret time travellers, not because they won't exist at some point in the future, but because they don't exist yet. More worryingly perhaps, if there is a moment in time that is as far back as you can travel, surely everyone will try to visit it at some point? There's a chance that if multiple people try to emerge at the exact same point in spacetime, they'll collide at the atomic level and annihilate themselves in a flash of radiation. Of course, if there are aliens who were making time machines at the time of the Roman Empire, and a couple of teleporters too, perhaps we could ask to borrow them. Although they haven't visited either as far as we know, unless the Romans forgot to mention it. And thus, you can see why it is easy to get lost in hypotheticals.

Engineering difficulties aside, is this sort of thing actually possible? 'Perhaps' is the only answer I have right now. These are mathematically interesting consequences of our framework for understanding the Universe (and even then there are different interpretations), but we're humble enough to admit that we don't comprehend everything. The framework itself is not complete, such as when we consider the integration of relativity and quantum mechanics. Interesting ideas might hide in the gaps of our theories but it isn't guaranteed. My personal feeling is that a more comprehensive framework might throw up reasons as to why we can't actually have wormholes or time machines. This spoils a little bit of our fun, maybe, but we'll also be closer to being able to explain the entire Universe, which would be nice.

Some Conclusions

Given the subject matter, it is traditional to end things on an uplifting note. Relating it all to an everyday experience is useful too and more comforting than dealing with the immensity of spacetime.

Let's try the other way.

Consider the far future, and I mean the far future. The Sun has become a red giant and then transformed into a planetary nebula, but even that is a distant memory now. Material from that nebula found itself part of new star-formation processes and new generations of stars were born and died. The Milky Way galaxy and the Andromeda galaxy crashed together and merged, and there are perhaps other collisions long forgotten too.

Entropy builds: high-energy volumes are transferred to lower-energy volumes. The Universe is cooling down. A population of black holes exists and feeds on anything that gets close enough, but it's getting increasingly difficult to find pockets of gas that collapse under gravity to undergo nuclear fusion and give birth to new stars.

So it continues. Throughout, the expansion of the Universe tends to move galaxies away from each other. On average, things are getting cooler, darker and increasingly far apart. We're approaching an equilibrium where nothing happens. Nothing can happen, as there are no concentrations of energy dense enough.

Black holes haunt this universal graveyard, and indeed may be the last interesting things remaining. As we've learned, they are the end results of various events. They interact, and even change to some extent, but they don't transform into something new. There will have been vast numbers of collisions and mergers over the long history of the Universe, but as we rush through time, these events will cease.

The very final act, observed by no one, is the slow evaporation of those black holes into a fizz of particles too lethargic and isolated to do anything. Perhaps these also decay further, or perhaps some quantum processes and grand cosmological framework reveal this is merely part of a larger multiversal story, but there are no witnesses to that.

Our Universe is, in some sense, at peace. Cold, quiet, dark and empty. The end.

I'm not 100% sure what I was going for there, but it ended up being pretty bleak. The problem is our current understanding of black holes and cosmology do suggest that this is the eventual fate of our Universe.

Thankfully, there are a few things with which we can console ourselves. The first is, of course, that the timescales involved are so enormous that there is literally no point worrying about this sort of thing. And, related to this, you certainly can't do anything about it! More interestingly though, we don't know everything and science has a long history of radically overhauling what we thought we understood about the Universe. Though we can't predict when they'll come or what the outcome would be, we might wager that there are still scientific revolutions of some sort in our future that will upturn our current comprehension of things. Maybe we'll see a change as meaningful as Einstein's improvements of Newton's work or a beautiful explanation like Schwarzschild's!

Even correctly predicting the origin of a revolution is perhaps unlikely. Nevertheless, if you were forced to gamble, the study of black holes would have a lot to recommend it. We've learned that these objects are created in fascinating circumstances and give rise to what seem like bizarre phenomena. Black holes involve velocities, densities and other properties at the limits of our current understanding and many are far beyond our abilities to recreate or influence directly. We can admit that we don't have a complete understanding of them. The puzzle is not completely solved.

Glossary

accretion disc – a dynamic disc-like structure of material spiralling towards a massive object such as a black hole. Things like dust and gas molecules colliding or rubbing against each other as they spiral inwards might emit radiation, which can then be used to map out the spacetime near a black hole's event horizon, though not inside it.

Chandrasekhar limit – the maximum mass of a white dwarf star, around 1.4 times the mass of the Sun. Beyond this, the gravitational attraction of the star's own matter cannot be balanced and further collapse to a neutron star or black hole is assured.

corpuscular theory of light – a theory of light proposed by Isaac Newton. It stated that light was made up of tiny particles called corpuscles that travelled in straight lines at high velocities. This speed could be changed by the medium through which they were travelling and as they impacted on the retina of the eye they would produce an image of the emitting source. Importantly, his theory was later disproved.

corpuscle – one of the tiny particles – assumed to have negligible mass, but differing sizes that would produce different colours – that Isaac Newton believed conveyed vision. Like the theory they relate to (see above), their existence has been debunked.

curvature – a measure of the warping of spacetime. Paths through 'curved' spacetime will deviate from that of a 'flat' spacetime. In spacetime, time itself is subject to curvature, so objects may move through time at different (and changeable) rates as well as in usual three-dimensional motion.

electron degeneracy pressure – an effect that resists compressing atomic matter arbitrarily densely. It prevents electrons orbiting an atomic nucleus from combining with the nucleus itself. If overcome, we create electron-degenerate matter such as that which makes up white dwarf stars.

entropy – a measure of the unavailability of thermal energy for the purposes of conversion into mechanical work. As entropy increases, availability decreases.

ergosphere – a region, beyond the event horizon, around a rotating black hole. Its shape and extent is dependent on the rotation and mass of the black hole.

escape velocity – the instantaneous velocity required for an object to escape the gravitational field of a massive object (such that the former will eventually reach infinite distance from the latter). It changes depending on the distance already travelled from the massive object and the mass involved. Rockets do not reach escape velocity to get into orbit, because they use fuel in addition to the velocity given to provide acceleration during ascent.

event horizon – the boundary beyond which gravity is too strong for even light to escape. The point of no return for infalling objects, beyond which direction becomes somewhat meaningless other than 'further in' towards the singularity at the heart of the black hole.

field equations – Einstein's mathematical expressions relating the distribution of matter within a spacetime and its geometry. At low velocities and in weak gravitational fields, they are much the same as Newton's law of gravitation, but deviate from it in strong gravitational fields or at high velocities. Full solutions are often impossible in anything but the simplest of situations.

force – an influence that alters the motion of an object. Unopposed, a force can cause a mass to accelerate; for a set mass, the acceleration is proportional to the size of the force.

gravitational acceleration – the increase in speed caused by the attractive force of gravity. According to Newtonian mechanics, this is a force that depends on the mass of the attracting body and the separation of the masses involved. On Earth, gravitational acceleration is approximately 9.8 m/s^2.

gravitational redshift – describes the process by which photons have to 'climb out' of a massive object's gravitational well, losing energy and shifting to longer wavelengths. Thus, light emitted from an object will be received at different wavelengths (or frequencies) depending on the receiver's position relative to the emitter.

Innermost Stable Circular Orbit (ISCO) – objects cannot maintain stable orbits at every distance from a black hole. The ISCO boundary marks the threshold at which orbits become unstable and objects move on a trajectory further towards the black hole. The position of the boundary depends on the black hole; for a non-spinning black hole it is a distance of three Schwarzschild radii.

kinetic energy – the energy of an object due to its motion. It is the amount of work required to accelerate an object from rest to a given velocity.

Lense–Thirring effect – near a rotating body with mass, spacetime itself is 'dragged' around. This causes the rotational axes of nearby masses to change and prevents those masses from maintaining a fixed orientation unless the dragging is counteracted, even if we ignore things like other orbital motions.

mass – a measure of the strength of a physical body's gravitational attraction to other bodies, and its resistance to acceleration by a force. Colloquially used interchangeably with weight, the latter is in fact the force acting on a body due to gravity. An object will have the same mass on the Earth and on the Moon, but will weigh less on the Moon due to a lower gravitational field strength there.

matter – a physical substance, taking up a volume of space and having mass. There are different ideas as to what is regarded as matter, but generally we consider it to be material composed of atoms, which are in turn composed of subatomic particles. Here, we mention subatomic particles like electrons, photons and neutrons, though these are themselves made of (or belong to classes of) elementary particles.

neutron degeneracy pressure – a quantum effect that resists compressing atomic matter arbitrarily densely. It prevents electrons and protons in electron-degenerate matter from combining inside the nucleus of atoms to form neutron-degenerate matter. Beyond the limits of neutron degeneracy pressure, collapse to a black hole occurs.

perihelion – the nearest point of an object's orbit (aphelion is the furthest point).

photon – discrete units of energy that make up what we think of as light. These elementary particles exhibit both particle-like and wave-like behaviour (wave–particle duality). They have no mass and travel at the speed of light (in empty space, this is approximately 300,000 kilometres per second).

photon horizon – the region around the black hole where photons have to travel in orbits, also known as the 'photon sphere' or 'photon circle'. Further out than this, they can potentially escape, but any closer in and they will move inevitably towards the black hole.

post-Newtonian approximations – expansions to Newton's law of gravitation when analytical solutions to Einstein's field equations aren't possible (which is usually the case). Higher-order approximations increase accuracy and can account for a number of relativistic influences, at the cost of increasing mathematical complexity.

potential energy – in this context refers to the energy held by an object with mass because of its position relative to another mass. Lifting a bag of sugar off the ground will increase its potential energy because of the gravitational attraction of the Earth. Lowering it again will reduce this potential.

progenitor star – a star that provides the origin of a phenomenon such as a supernova, which may significantly alter the star's size, temperature, mass, composition or structure.

proper distance – the invariant measure of the separation between two 'events' in spacetime, independent of observers. This accounts for intervals of separation in space, but also in time (it is simply another dimension within spacetime). It seems complex, but we are often considering different 'points of view' within a spacetime that is distorted by strong gravitational fields. We need measurements independent of one particular position within spacetime.

Schwarzschild radius – the mass-dependent extent of the event horizon of a Schwarzschild black hole. While not quite right to call it the 'size' of the black hole, it is nevertheless a characteristic size associated with this class of object and often informally used as such.

singularity – a point that is not mathematically defined (points in an equation heading towards infinity, for example). With regard to black holes, these are the Schwarzschild radius and the centre of the black hole.

spacetime – a four-dimensional geometry describing the Universe, in which the three dimensions of space are merged with the dimension time. An object occupies a position within spacetime. It may help to consider motion – that is, an object may move in three-dimensional space at different velocities, but also through time at different rates.

spaghettification – a tidal effect caused by strong gravitational fields. When falling towards a black hole, for example, an object is stretched in the direction of the black hole (and compressed perpendicular to it as it falls). In effect, the object can be distorted into a long, thin version of its undistorted shape, as though being stretched like spaghetti.

stellar mass – a term referring to concentrations of mass approximately equal to that of the Sun (referred to as one solar mass). Like the term 'supermassive', it is somewhat vague and may even be employed to include descriptions involving thousands of solar masses.

supermassive – usually used in reference to masses in the order of millions or billions of solar masses.

supernova – an enormously violent and luminous stellar explosion. There are different supernova types and physical mechanisms responsible for the event, but in essence the star explodes, throwing off its outer layers. The inner layers collapse and depending on their mass may create objects such as neutron stars or black holes.

time dilation – the difference in elapsing time, as measured by two clocks, primarily due to differences in velocity or the gravitational fields the clocks are in. A measure of the difference in the passage of time itself in various circumstances.

Tolman–Oppenheimer–Volkoff limit – the maximum mass a neutron star can have before it collapses into a black hole. The precise limit depends on whether the star is spinning and how fast, but it is roughly between two and three solar masses.

velocity – a measure of speed and direction (in other words a measure of the rate of change of position with respect to a frame of reference as a function of time). Colloquially, speed and velocity are often used interchangeably, but in many situations the direction of motion is a key issue so the term should be used with care.

Royal Observatory
Greenwich Illuminates

Stars
by Dr Greg Brown
978-1-906367-81-7

Planets
by Dr Emily Drabek-Maunder
978-1-906367-82-4

The Sun
by Brendan Owens
978-1-906367-86-2

Space Exploration
by Dhara Patel
978-1-906367-88-6